MAGGIE'S
Double Bluff
DISCOVERIES

written by Amanda Brager
photos by Pam Brager

Manon,
I hope you and
your brother have
as much fun as
maggie and Drover do!

much love,
from
maggie + Amanda

MINDSTIR MEDIA

Published by Mindstir Media, LLC
1931 Woodbury Ave. #182 | Portsmouth, New Hampshire 03801 | USA
1.800.767.0531 | www.mindstirmedia.com

Printed in the United States of America
ISBN-13: 978-0-9972233-6-1
Library of Congress Control Number: 2016902100

DEDICATED TO PROFESSOR FRANK CAMPBELL
WHO TAUGHT ME EVERYTHING I KNOW ABOUT SCIENCE.

When my brother Ben was 12, he was diagnosed with a type of vasculitis called Granulomatosis with Polyangiitis (Wegener's). Drover came to cheer Ben up in 2004 and twelve years later is still doing that job very well!

A percentage from every book sold will be given to Ben Brager's Medical Fund to help him with lost wages and medical expenses. If you would like to contribute directly, please contact any Wells Fargo Bank.

Spending time at the beach with Maggie and Drover is a great way to relax.
I hope you enjoy your "visit!"

Amanda Brager

MAGGIE'S
Double Bluff
DISCOVERIES

DROVER IS AN ALL WHITE
PARSON RUSSELL TERRIER.

MAGGIE IS HIS NIECE.

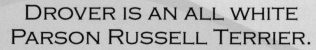

Maggie quivered with excitement as she waited to board the ferry. She was on her way to visit her Uncle Drover who lives on Whidbey Island in Washington. As she waits for the ferry she remembers listening to her uncle tell her all about the island. He told her that it is 62 miles long, making it the longest island in the United States.

Maggie could hardly stand the excitement. Drover always took her to fun places. Maggie wondered where they would be going this time. Before she knew it, she had arrived at Drover's house. He told her, "We're going to Double Bluff!"

When they arrived at the beach, Drover pointed out the hazardous bluff sign. He told Maggie to be very careful because the bluffs were formed by glacial till making them unsafe to climb.

"Uhmm, Uncle Drover? What is glacial till?" Maggie questioned.

Drover explained that it is a bunch of rocks and finely ground material that were carried by a glacier and then dumped when the ice melted. Maggie stood in awe hardly believing that. "All this came from a glacier?"

"Yup. A long time ago. Are you ready to get down to the beach?" Drover asked. In answer, Maggie scampered down the stairs to the sandy beach.

Drover caught up with her and hopped up on a rock. "Look at all the life around you, Maggie."

She spun around in a circle. "Uhmm, Uncle Drover?" Maggie said in a questioning tone. "I only see a couple people."

"There is so much life on the beach. If you look closely, you can see that all kinds of interesting animals live out here. Come on I'll show you."

Drover hopped off of the rock and headed down the beach. Maggie followed behind him, wondering about the animals they would find.

"UNCLE DROVER!" Maggie squealed with excitement. "Come look at this tide pool!" As Drover headed toward Maggie, she headed into the pool to explore.

A few seconds later Drover heard a scream. He ran to Maggie's side wondering what was wrong. "Uncle Drover! What IS that?" Maggie pointed to a strange looking orange tube.

Drover examined the tube. "That's a sea cucumber, Maggie. It's a filter feeder." Maggie looked confused.

"Filter feeding animals live underwater most of the time and feed on food that is just floating by. They use their feather-like feet to collect their food and bring it to their mouths."

"Sounds weird," Maggie said as she dashed off.

Drover trolled along behind Maggie looking for more life to show to her. He spotted something and called Maggie over. She turned around and headed to her uncle.

Drover pointed to a red animal shell and asked Maggie, "Do you know what this is?"

"That's a lobster, right?"

"Not quite. It's a crab," Drover corrected.

Maggie looked worried. "Uhmm, Uncle Drover? Won't it pinch me?"

Drover checked it out and saw that it was an empty shell. He told her how crabs molt, leaving their old shell behind and growing a new shell. Knowing that, Maggie crept toward the crab shell. "OH, Uncle Drover! It smells absolutely wonderful!" Maggie giggled. "Can I roll in it? Please?"

Drover thought about it for a while and decided they should find something even smellier.

"Okay!" Maggie squealed with excitement as she bolted down the beach.

Maggie stopped suddenly. "Uhmm, Uncle Drover?
Why are those people digging holes in the middle
of the beach? Won't someone step in them and get
hurt?"

Drover explained that the people were digging for
clams. "Come here and I'll show you how to dig
clams. And I'll show you how to fill your holes in
so it's not dangerous."

Drover saw a clam squirt and started digging. He dug and dug and clams shot out behind him. Maggie examined them and wondered if she could eat them. Drover explained that clams need to be cooked before being eaten.

"Yum, sounds tasty," Maggie said. She then scampered out to the water.

Maggie put one foot in the water and squealed. Then she yelled in a worried tone, "Uhmm, Uncle Drover?" She backed away from the water. "What is all this slimy stuff? Is it some kind of animal, too? Will it bite me?"

Drover snickered and looked at the seaweed Maggie was afraid of. He walked into the water and told her that it was just seaweed. "It's a plant, it can't hurt you."

"Are you sure?" questioned Maggie as she once again eased towards the red, green, and brown slimy plants.

"It's harmless. Lots of animals eat seaweed and use it for protection. Some people even eat it. The best part about seaweed is it's great to roll in!" They enjoyed a nice long roll in the smelly, rotting seaweed and then continued along the beach.

Drover was looking at something in shallow water when Maggie walked over to see what he was checking out. "Eww! What is that? Is it some kind of snake?"

Drover explained it was not an animal, but a plant called bull kelp. He told her it can grow up to two feet per day right here in the Pacific Northwest.

"Wow! That's a lot! What else lives around here, Uncle Drover?" Maggie asked, feeling more confident.

Drover encouraged Maggie to continue down the beach to see some more amazing sea creatures.

"All right!" yelled Maggie as she caught up with Drover.

Maggie stopped running and started hopping around. "OUCH!" she squealed. "I just stepped on something sharp." She sniffed the area, trying to find what hurt her. Drover arrived and Maggie pointed at something. "Right there, Uncle Drover. What is that?"

"That's a barnacle, Maggie. They have to be tough so they don't get hurt when the tide goes out and they are stranded out of water."

"Well, my foot really hurts." Maggie complained.

Drover decided it was a good time to head home. Maggie limped along behind him.

A few minutes later Maggie had forgotten all about her injury. She was ready to explore some more. She stumbled upon a large shell and picked it up. "What is this, Uncle Drover?"

Drover explained that the beautiful swirled shell was a moon snail.

"Oh, cool," Maggie answered, not asking any questions because she found something else to sniff.

When she looked to her left she noticed a strange lump
of sand. "What it THIS?" she wondered out loud.

Drover began to explain, "It's a moon snail collar, it..."

Maggie interrupted, "Uhmm, Uncle Drover, this collar
doesn't fit," she complained.

Drover turned around and found Maggie doing
something strange. "Actually," he said trying to hide a
smile, "that's not supposed to be worn around your
neck. It's how moon snails lay their eggs."

"So I can't wear it?" Maggie asked.

"No, it will fall apart when it dries out. Why don't you put it back so the snails can hatch."

Maggie and Drover slowly walked back along the water line. They sat down to enjoy the beach just a little more before they left. Out of the corner of his eye Drover noticed his favorite sea creature.

He got up and Maggie followed. "What are you looking at Uncle Drover?"

"Actually," he responded, "you're stepping on it."

"OHH!" Maggie squeaked. "Is that a sea star? Will it bite me?" Drover confirmed that it was a sea star but explained that it ate things like clams not dogs. He told her that sea stars were his favorite animal on the beach. "It's so bright!" She noticed.

"It's beautiful. Isn't it?" Drover asked. Maggie agreed. Drover told Maggie that they were his favorite animals because of all the different colors. He then told her that he has seen blue, orange, purple, and yellow sea stars. Maggie was amazed and decided sea stars would be her favorite animal also.

They walked the long mile back to Drover's house. They played for a while after dinner and talked about their adventures. By this time it was dark. "Maggie, look up," Drover said as he pointed toward the night sky.

In a slow, tired voice Maggie said, "It's the Big Dipper."

"Do you know the story of the Big Dipper?" Drover wondered. Maggie didn't respond. Drover looked over to find Maggie asleep. He chuckled quietly, "Good night, Maggie." He tucked Maggie into bed and went to sleep and dreamed about sea stars.